Vertebrates from Island of Tamaulipas, Mexico

Richard F. Johnston, Gerald G. Raun

Robert K. Selander, B. J. Wilks

Alpha Editions

This edition published in 2024

ISBN : 9789362923073

Design and Setting By
Alpha Editions
www.alphaedis.com
Email - info@alphaedis.com

As per information held with us this book is in Public Domain.
This book is a reproduction of an important historical work. Alpha Editions uses the best technology to reproduce historical work in the same manner it was first published to preserve its original nature. Any marks or number seen are left intentionally to preserve its true form.

The Ecological Setting

The barrier island of Tamaulipas geologically and ecologically resembles Padre Island, of the Gulf coast of lower Texas, north of the mouth and delta of the Rio Grande. South of the delta, the island in Tamaulipas is a narrow strip of sand less than a mile in average width and is broken by a series of narrow inlets or "passes" through which water from the Gulf of Mexico mingles with that of the Laguna Madre de Tamaulipas. The passes are subject to recurrent opening and closing. North of the mouth of the Río Soto la Marina, eight passes are designated by local fishermen, but only three, the Third, Fourth, and Fifth, were open at the time of our visit.

The Laguna Madre de Tamaulipas is described by Hildebrand (1958) in connection with a preliminary study of the fishes and invertebrates there. The average depth is probably less than 70 cm. and the waters are hypersaline. In the time of the recent drought in Texas and northeastern México, salinity varied from 108 to 117 parts per thousand in the northern part of the laguna near Arroyo del Tigre (measurements taken in March, 1955) to from 39 to 48 parts per thousand in the southern part near Punta Piedras (measurements taken in October and November, 1953, and in March, 1954). Discussions of the geologic history, ecology, and zoogeography of the lagoons of the Gulf coast of the United States are given by Hedgpeth (1947; 1953).

Localities in coastal Tamaulipas mentioned in the text of this paper are shown on Plate 5.

The principal animal habitats are found in three vegetational associations (plates 6 and 7). On flats and low dunes lying between, and partly sheltered by, larger active dunes, small clumps of *Croton punctatus* and a sedge (*Fimbristylis castanea*) are the only conspicuous plants. Near the western edge of the dunes, *Ipomoea pescaprae* var. *emarginata* is mixed with *Croton*, and there are scattered clumps of shrubby wolf-berry (*Lycium carolinianum* var. *quadrifidum*), and mesquite (*Prosopis juliflora*).

The dunes are relatively stabilized on the western side of the island, and there we found moderately dense stands of mesquite trees reaching heights of from eight to 10 feet. Prickly-pear cactus (*Opuntia lindheimeri*) was common in those stands of mesquite, and we saw an occasional yucca tree. A fairly dense ground cover was formed by blanket-flower (*Gaillardia pulchella*), marsh-elder (*Iva* sp.), *Flaveria oppositifolia*, *Enstoma exaltatum*, and *Croton capitatus* var. *albinoides*.

A more open, xeric expression of the mesquite-cactus vegetation occurs on exposed, low clay dunes (see description by Price, 1933) located on alkaline flats bordering the laguna. At the time of our visit, most of the mesquites in these stands were dead or dying, the cactus was abundant, and the ground cover, which was sparse, included drop-seed (*Sporobolus virginicus*), ragweed (*Ambrosia psilostachya*), and *Commicarpus scandens*.

On alkaline flats flooded by hypersaline waters of the laguna following heavy rains, *Batis maritima* is found in the lower areas, but on the slightly elevated areas there is low and almost continuous cover of *Monanthochloë littoralis*, in which can be found *Batis*, *Borrichia fructescens*, *Salicornia* sp., *Iva* sp., and sea-lavender (*Limonium carolinianum*).

Near Third Pass, sea oats (*Uniola paniculata*), evening primrose (*Oenothera* sp.), and cordgrass (*Spartina* sp.) are present on the dunes, and on alkaline flats we collected *Conocarpus erectus*, *Leucaena* sp., and *Cassia fasciculata* var. *ferrisiae*.

Itinerary

We reached Washington Beach from Matamoros on July 6, and drove to a point approximately 33 miles south on the beach, where we made Camp 1 on the east side of large dunes 400 yards from the surf. From this camp we worked the beach and dunes and also visited alkaline flats adjacent to the Laguna Madre. On the afternoon of July 8, we drove south along the beach and established Camp 2 on the south side of the Third Pass, approximately 73 miles south of Washington Beach. We had intended to go farther south but were unable to cross the Fourth Pass, an inlet three miles south of the Third Pass. We left the barrier island on the afternoon of July 10, after driving north from Camp 2 to the mouth of the Rio Grande, 11 miles north of Washington Beach.

Mexican fishermen camped at the Fourth Pass told us that, had we been able to cross the Fourth Pass, it would have been possible to drive south on the beach all the way to La Pesca, a fishing village near the mouth of the Río Soto la Marina, approximately 150 miles south of Washington Beach.

Summary of Previous Work in the Area

The ornithologist H. E. Dresser (1865-1866) worked in southern Texas and at Matamoros, Tamaulipas, in 1863, and on one occasion reached the mouth of the Rio Grande ("Boca Grande"). He did not visit the barrier island or the Laguna Madre de Tamaulipas.

In their extensive travels through México, E. W. Nelson and E. A. Goldman made collections at three localities in the coastal region of Tamaulipas but did not reach the barrier island (Goldman, 1951). Goldman collected at Altamira, near Tampico, from April 2 to 24, 1898, and from May 15 to 20 of the same year both he and Nelson made headquarters at Altamira. Nelson and Goldman also collected in the vicinity of Soto la Marina, 25 miles from the coast, from March 1 to 10, 1902, and, from February 13 to 15, they visited Bagdad, described by Goldman (1951:260) as "a village at very low elevation on the Río Grande about 6 miles above the mouth of the river."

In March, 1950, C. von Wedel and E. R. Hall collected four species of mammals and one bird on the barrier island at Boca Jésus María (Eighth Pass). A report of this work published by Hall (1951) contains descriptions of three new subspecies of mammals from the island.

A few records of birds from the southern end of the barrier island and from other parts of coastal Tamaulipas were reported by Robins, Martin, and Heed (1951). In 1953, R. R. Graber and J. W. Graber made ornithological studies in the vicinity of Tampico and also reached the western edge of the Laguna Madre de Tamaulipas. Several papers on this work have appeared (Graber and Graber, 1954a, 1954b; Graber, 1955), but a comprehensive account of their observations and specimens was not published. Finally, J. R. Alcorn collected some sandpipers 20 miles southeast of Matamoros, on August 21, 1954, obtaining the first record of the Semipalmated Sandpiper (*Ereunetes pusillus*) in Tamaulipas (Thompson, 1958).

Accounts of Species

Catalogue numbers in the following accounts are those of the Museum of Natural History, The University of Kansas.

Reptiles

Gopherus berlandieri Agassiz: Texas Tortoise.—A pelvic girdle and complete shell with a few attached scutes (63494) were found in stabilized dunes at Camp 1 on July 7, and tracks were seen in the same area. Fragments of two other shells (63493, 63495) were found on sand flats between active dunes at Camp 1.

Holbrookia propinqua propinqua Baird and Girard: Keeled Earless Lizard.—This lizard was abundant on dunes and in pebble-strewn blow-out areas between dunes at Camp 2, but it occurred in smaller numbers in the less stabilized dunes of sparser vegetation at Camp 1. Breeding was in progress at both localities, as evidenced by the presence of eggs in the oviducts of several females, by the heightened coloration of both sexes, and by mating behavior.

The mating behavior of this species has not been described in the literature, and the following observations, made by Raun at Camp 2 on July 8, may be of interest. A male was seen to circle a female as the latter remained motionless with tail curved upward and to the side, exposing a patch of bright pink-orange color on the ventral surface of the tail. At times the male approached the female from the rear and slightly to the side, biting the dorsal part of her neck and simultaneously attempting to effect intromission. The female several times reacted to this approach by running forward a few steps, thereby freeing her neck from the grasp of the male. When the male did not attempt to approach again, the female appeared to invite copulation by moving in front of him with tail elevated and the colored ventral surface prominently displayed. At the time of copulation, the male mounted from the rear on the right side of the female, grasped her neck, and circled his tail beneath her tail; at the same time the hindquarters of the female were arched upward.

To confirm the presumed sexes of the two individuals under observation, both were collected while in copulation. Examination of the still-coupled specimens showed that both hemipenes of the male were everted and the left one had been inserted.

Apparently the pink-orange subcaudal patch of females is present only in the mating season. It was not present on specimens of this species taken by Raun and Wilks on Padre Island, Texas, in autumn, and it is not mentioned in taxonomic descriptions by Axtell (1954) and Smith (1946).

Measurements of adult specimens in our series indicate that females are of smaller average size than males, and, as previously noted by Smith (1946:132), females of this species have disproportionately shorter tails than do males (Table 1).

Holbrookia propinqua was previously collected on the barrier island by Axtell (1954:31; see also Axtell and Wasserman, 1953:2), who took specimens at Boca Jésus María, at a locality six to seven miles south of Boca Jésus María, and at a point 20 miles east-southeast of Matamoros. Axtell (*loc. cit.*) also lists specimens in the Museum of Zoology, University of Michigan, from Tepehuaje and from one mile north of Miramar Beach (Tampico).

Specimens (56): 3 ♂ ♂ adult, 1 ♂ subadult, 63433-436, Camp 1, July 7. 33 ♂ ♂ adult, 63437-440, 63443-445, 63447, 63448, 63450-456, 63458, 63460, 63462, 63463, 63465-468, 63470-478; 13 ♀ ♀ adult, 63441, 63446, 63449, 63457, 63459, 63469, 63479-485; 6 juv., 63442, 63461, 63464, 63486-488; Camp 2, July 9-July 10.

TABLE 1.—MEASUREMENTS IN MILLIMETERS OF ADULT SPECIMENS OF *Holbrookia propinqua* FROM THE BARRIER ISLAND OF TAMAULIPAS

Sex	Number of specimens	Snout-vent length	Tail length	Ratio: snout-vent to tail
Male	33	56.0±0.5[A] (49-62)	77.0±0.7 (69-85)	0.731±0.001 (0.682-0.817)
Female	14	50.9±0.5 (47-53)	62.2±0.9 (57-68)	0.825±0.001 (0.735-0.877)

[A] Mean ± standard error; range indicated in parentheses.

Cnemidophorus gularis Baird and Girard: Whip-tailed Lizard.—At both camps we found this species in the same general habitat in which *Holbrookia* occurred, but in numbers decidedly fewer than the latter.

Specimens (4): 2 ♀♀ adult, 63489, 63490, Camp 1, July 7. 1 ♂ adult, 63491, 1 ♀ adult, 63492, Camp 2, July 9.

We failed to take specimens of snakes on the barrier island, but tracks of snakes were noted on two occasions in dunes near Camp 1; one trail led into a burrow of a kangaroo rat.

Birds

Unless otherwise indicated, specimens taken were not molting. For birds undergoing postnuptial or postjuvenal molt, the degree of advancement of the molt is indicated by recording the number of primaries of the old plumage that have not been dropped. For example, the designation "4 P old" signifies that all primaries except the distal four have been molted.

Table 2 presents results of a strip census of birds along the strand, made by three of us from the moving truck on the morning of July 10. Birds characteristically found on sand near the surf were thus conveniently counted in accurate fashion. Birds not ordinarily found on the strand could not be treated this way; most were considerably less abundant than the eight most numerous species listed in Table 2. Over-all, the numbers of individuals listed are a good index of abundance of the Great Blue Heron and of the common charadriiform birds on the beach in early July. The Black Tern is an exception, however, and this is discussed in the account of that species on page 327.

TABLE 2.—BIRDS[A] RECORDED ALONG 17 MILES OF BEACH[B] ON THE BARRIER ISLAND OF TAMAULIPAS

Species	Number	Birds per mile
Great Blue Heron	9	0.5
Oyster-catcher	1	0.1
Black-bellied Plover	20	1.2
Wilson Plover	53	3.1
Willet	43	2.5
Sanderling	55	3.2
Laughing Gull	136	8.0
Black Tern	19	1.1
Caspian Tern	82	4.8
Least Tern	221	13.0
Royal Tern	301	17.7
Cabot Tern	122	7.2
	Total: 1062	Total: 62.4

[A] Common Tern, Forster Tern, and Long-billed Curlew also seen but not counted.

[B] Between 56 and 73 miles south of Washington Beach, 11:00 to 11:45 a. m., July 10, 1961.

Pelecanus erythrorhynchus Gmelin: American White Pelican.—A flock of approximately 300 individuals was seen resting at the edge of the Laguna Madre near Camp 2 on July 9. When disturbed by gunshots, the birds circled high over the laguna and flew to the west. Among bones found on sand flats at Camp 1 are a left tarsometatarsus and a pedal phalanx of an American White Pelican.

Supposedly the only breeding colony of this species on the northern Gulf coast is one in the Laguna Madre near Corpus Christi (Peterson, 1960:8), but the possibility of one or more such colonies existing in northeastern Tamaulipas has been suggested by Amadon and Eckelberry (1955:68) on the basis of their observations of individuals seen soaring near the coast 15 to 20 miles south of Brownsville on April 15 and June 5, 1952. According to Hildebrand (1958:153, and personal communication, August 14, 1961), small colonies of white pelicans do breed in some years on two small islands, in the Laguna Madre of Tamaulipas, located at 25° 26´ North and 93° 30´ West.

In Veracruz the species is recorded as a winter visitor and transient (Loetscher, 1952:22; Amadon and Eckelberry, 1955:68). Coffey (1960:289) reports the following observations for Veracruz and Tamaulipas: a flock of 52 between Tlacotalpan and Alvarado, May 29, 1951; 80 near Cacaliloa, April 20, 1958; 180 birds north of Alvarado, April 24, 1958; four at Altamira, May 28, 1955; flocks of three, 13, and 37 "south" of Matamoros, May 20, 1951; 72 at Lomas del Real, November 20, 1956.

Pelecanus occidentalis Gmelin: Brown Pelican.—Three individuals flew north over the surf near Camp 1 on July 7, and a lone bird was seen diving into the Gulf a short distance beyond the surf near Camp 2 on July 9. Birds seen by us probably were of the population named *P. o. carolinensis*, which is resident along the Gulf coast (Mexican Check-list, 1950:21).

Phalacrocorax sp.: Cormorant.—From 80 to 100 adult and juvenal cormorants were on the laguna at Camp 2 on July 8 and 9. Probably they were Common Cormorants (*P. olivaceus*), but, because specimens were not taken, we cannot eliminate the possibility that some (or all) were Double-crested Cormorants (*P. auritus*). The former breeds in coastal lowlands of eastern México, whereas the latter is known in eastern México only as a winter visitant and has not been recorded in Tamaulipas (Mexican Check-list, 1950:24).

Fregata magnificens Mathews: Magnificent Man-o'-war Bird.—An observation of a lone bird circling high over the laguna at Camp 2 on July 9 seemingly constitutes the third record of this species in Tamaulipas. Previous records were reported by Robins, Martin, and Heed (1951:336), who found "large numbers" in the Barra Trinidad region (8 miles north of Morón) on April 27 to 29, 1949, and mentioned an immature male taken at Tampico on

April 23, 1923; this specimen has been identified by P. Brodkorb as *F. m. rothschildi*.

Ardea herodias Linnaeus: Great Blue Heron.—Our records of this heron are limited to the following observations: four individuals on the beach and seven in the laguna at Camp 1, July 7; one on the beach 52 miles south of Washington Beach, July 8; one 74 miles south of Washington Beach, July 8; two at Third Pass, July 8; 41 standing on mud-flats at the edge of the laguna near Camp 2, July 9; nine on the beach 56 to 73 miles south of Washington Beach, July 10; one on the beach 42 miles south of Washington Beach, July 10.

The status of the Great Blue Heron in coastal Tamaulipas remains to be determined. The subspecies *A. h. wardi* (considered a synonym of *A. h. occidentalis* by Hellmayr and Conover, 1948) is resident and breeds on the Gulf coast of Texas and is to be expected as a resident in Tamaulipas (Mexican Check-list, 1950:27). The species may breed south to Veracruz, where Loetscher (1955:22) reports it is "regular at nearly all seasons, chiefly on the coastal plain"; he records an observation near Tamós on July 1. The subspecies *A. h. herodias* and *A. h. treganzai* winter through much of México and have been recorded in Tamaulipas (Mexican Check-list, 1950:27).

Florida caerulea (Linnaeus): Little Blue Heron.—We saw a white (immature) individual feeding with Reddish Egrets along an inlet at Camp 2 on July 8.

Dichromanassa rufescens rufescens (Gmelin): Reddish Egret.—This egret was recorded only about the inlet at Camp 2, where 15 individuals were feeding, either singly or in small groups, on July 8 and 9. We noted frequent use of the "Open Wing" method of foraging, as described by Meyerrieks (1960:108).

Specimen: ♀ juv., 38899, ovary inactive, 587 gm., Camp 2, July 8. This specimen is referable to the nominate subspecies, which is resident along the Gulf coast. Our record seems to be the first for the species in Tamaulipas.

Leucophoyx thula (Molina): Snowy Egret.—Ten individuals of this species were feeding in association with Reddish Egrets in the inlet at Camp 2 on July 9.

Hydranassa tricolor (P. L. S. Müller): Tricolored Heron.—An observation of one individual flying along the margin of the laguna near Camp 2 is our only record of this species.

Nycticorax nycticorax (Linnaeus): Black-crowned Night Heron.—This heron was found only at the edge of the laguna near Camp 2; ten individuals were noted on July 8, and 20 were seen perched in a clump of mesquite trees on July 9. Perhaps half the birds seen were in juvenal plumage. A juvenile was shot and examined on July 9 but was not preserved as a specimen.

There appears to be no definite evidence of breeding by this species in Tamaulipas (Mexican Check-list, 1950:32), but such may be expected because the species breeds locally in Texas (Peterson, 1960:19) and in Veracruz.

Ajaia ajaja (Linnaeus): Roseate Spoonbill.—On July 9 at Camp 2, 38 spoonbills flew up from the edge of the laguna where they had been resting near a large flock of white pelicans.

Cathartes aura (Linnaeus): Turkey Vulture.—One Turkey Vulture was seen flying east at a point 2 miles west of Washington Beach on July 10. It is noteworthy that we saw no Yellow-headed Vultures (*C. burrovianus*), a species recently recorded in the region of Tampico north to Lomas del Real (Graber and Graber, 1954*a*).

Colinus virginianus texanus (Lawrence): Bob-white.—This species was seen only in or near clumps of mesquite near Camp 1, where three covies (7, 13, and 18 individuals) were flushed on July 7. Specimen: ♂ juv., 38900, testis 3 mm., 100 gm., 6 P old, Camp 1, July 7.

Porzana carolina (Linnaeus): Sora Rail.—On sand flats at Camp 1 we found a left humerus and several other post-cranial skeletal elements that have been identified by Dr. Pierce Brodkorb as belonging to this species. All the bones are of Recent age. We have no other record of the Sora Rail on the barrier island, but in all probability it occurs as a migrant and winter visitant along margins of the laguna.

Haematopus ostralegus Linnaeus: Oyster-catcher.—One individual was seen at Camp 2 on July 8, three were noted at the same locality on July 9, and one was present on the beach 72 miles south of Washington Beach on July 10. The only previous records of this species in Tamaulipas are a specimen (♂, 29348) taken by E. R. Hall 10 miles west and 88 miles south of Matamoros on March 20, 1950 (herewith reported for the first time), and three seen on the beach near Tepehuaje on May 9, 1949 (Robins, Martin, and Heed, 1951).

Squatarola squatarola (Linnaeus): Black-bellied Plover.—Plovers of this species were uncommon but regular on the beach; frequently two individuals were seen together, sometimes in association with one or more Willets. Specimens (4): ♂, 38915, testis 4 mm., 231 gm.; ♂, 38914, testis 4 mm., 221 gm.; ♂, 38916, testis 3 mm., 209 gm., Camp 1, July 7. Male, 38917, testis 4 mm., 186 gm., Camp 2, July 9. The specimens were molting (3-4 P old) into winter plumage and showed little or no subcutaneous fat.

Our specimens and records probably pertain to nonbreeding individuals summering on the coast, as the species is known to do in Texas (Hagar and Packard, 1952:9) and elsewhere in its range (Eisenmann, 1951:182; Haverschmidt, 1955:336; A.O.U. Check-list, 1957:174). In any event, our dates (July 6 to 10) are unusually early for autumnal migrants; they do not reach Texas until August (Peterson, 1960:94), and Loetscher (1955:26) gives August 7 as the earliest date for southbound migrants in Veracruz.

Charadrius hiaticula semipalmatus Bonaparte: Ringed Plover.—We have a single record, an adult male (38913, testis 2 × 1 mm., heavy fat, 47.0 gm., 4 P old) taken on a sandbar at Camp 2 on July 9. The bird was feeding in company with a flock of Sanderlings.

There is no previous record of the Ringed Plover in Tamaulipas. In Texas, Hagar and Packard (1952:8) indicate that the first autumnal migrants reach the central Gulf coast in the last week of July. In coastal México, the species has previously been recorded from August 23 to May 12 (Mexican Checklist, 1950:91). Therefore, the present record must represent an exceptionally early southbound migrant, or, more probably, a nonbreeding, summering individual. According to the A.O.U. Check-list (1957:166), nonbreeding birds are found in summer in coastal areas south to California, Panamá, and Florida. Many individuals spend the northern summer along the coast of Surinam (Haverschmidt, 1955:336).

Charadrius wilsonia wilsonia Ord: Wilson Plover.—This small plover breeds commonly on the beach and on alkaline flats adjacent to the laguna. Previous evidence of breeding in Tamaulipas consisted only of a report of a male with brood patches and an enlarged testis taken near Tamós on May 30, 1947 (Loetscher, 1955:26).

We saw many pairs of adults and a large number of well-grown juveniles, and, at a point 4 miles south of Washington Beach, we collected a brood of three small juveniles that had only recently hatched. The breeding season apparently was drawing to a close, for several adults in our collection were in postnuptial molt and showed marked gonadal regression. From July 6 to 9, a few small groups of birds were noted, but large groups were not seen until July 10, when several flocks of up to 60 individuals were found along the coast 3 to 7 miles south of Washington Beach.

Specimens (12): ♂, 38904, testis 4.5 × 2 mm., 58 gm., 3 P old, brood patches refeathering; ♂, 38905, testis 5 × 2 mm., 59 gm., 4 P old, brood patches refeathering; ♂ juv., 38903, 6.2 gm.; 2 sex?, 38901, 38902, 5.7 and 6.2 gm., 4 miles south of Washington Beach, July 6. Male, 38907, testis 5 × 2 mm., 56 gm., 7 P old, brood patches refeathering; ♀, 38906, ova to 1 mm., 61 gm., 3 P old, brood patches refeathering; ♀ juv., 38908, ovary inactive, 54 gm., in body molt; Camp 1, July 6. Male, 38910, testis 6 × 3 mm., 60 gm., 4 P old;

♀, 38909, ova to 1 mm., 57 gm., 4 P old, brood patches refeathering; Camp 1, July 8. Male, 38911, testis 2 × 1 mm., 55 gm.; juv., 38912, no weight or sex recorded; Camp 2, July 9.

Numenius americanus parvus Bishop: Long-billed Curlew.—Lone individuals and groups of two to five were noted occasionally along the beach each day. In total, some 30 to 50 birds were counted, but some individuals may have been recorded more than once on different days. Specimens (2): ♂, 38918, testis 4 mm., some fat, 459 gm., Camp 2, July 9; ♀, 38933, ova to 1 mm., no weight recorded, Camp 2, July 8.

Our assumption that some or all individuals seen by us were nonbreeding, summering birds is supported by the fact that our specimens are referable to the small, northwestern subspecies, *N. a. parvus*, rather than to *N. a. americanus*; the latter breeds south in the eastern United States to south-central Texas (A.O.U. Check-list, 1957:181). Loetscher (1955:27) saw a flock of 39 curlews near Tamós on June 30, and he notes that nonbreeding birds are fairly common at all seasons in Veracruz. Similarly, the species is present throughout the year on the central Gulf coast of Texas (Hagar and Packard, 1952:8). Authors of the Mexican Check-list (1950:94) do not mention the possibility that birds of this species recorded in México in July are summering rather than migrating. Twelve supposed migrants seen along Laguna Chila (Cacalilao), Veracruz, by Coffey (1960:291) on May 31, 1957, may have been summering birds.

Limosa fedoa (Linnaeus): Marbled Godwit.—Three were seen in shallow waters of the laguna at Camp 2 on July 9. Specimen: ♂, 38919, testis 6 × 2 mm., fat, 305 gm., 6 P old, Camp 2, July 9. Probably our records were of nonbreeding birds, which are known to occur in summer elsewhere in México (Mexican Check-list, 1950:94), sparingly in Texas (Hagar and Packard, 1952:8), and in South Carolina (A.O.U. Check-list, 1957:205). Apparently the only record for this species in Veracruz is one seen on May 11, 1954, east of Cacalilao (Coffey, 1960:292).

Tringa melanoleuca (Gmelin): Greater Yellowlegs.—Three birds were seen on alkaline flats at Camp 1 on July 7, and two were noted at Camp 2 on July 9. There is one previous report of this species in Tamaulipas, and, since it has been recorded as a migrant and winter resident in México between July 26 and April 26 (Mexican Check-list, 1950:95), our records seem to pertain to unusually early autumnal migrants or, possibly, to nonbreeding, summering birds. Other mid-summer records are available from Tamós on June 30 and July 1, and the species is "to be expected every month of the year" in Veracruz (Loetscher, 1955:27). Sight records for Veracruz in May (Coffey, 1960:291) may well pertain to summering birds. There are northern-summer records for this species from Texas (Hagar and Packard, 1952:8),

Surinam (Haverschmidt, 1955:367), and other areas within the winter range of this yellowlegs (A.O.U. Check-list, 1957:190).

Catoptrophorus semipalmatus semipalmatus Gmelin: Willet.—The Willet was common on the island. We found evidence of breeding and also saw large flocks of birds that were either nonbreeders summering in the area or early, postbreeding migrants from more northerly places. All along the beach and at the edge of the laguna at both camps we found Willets in twos or threes, often accompanied by one or two Black-bellied Plovers. On July 10 a small juvenile was captured; two adults in breeding plumage evidenced obvious concern at this action. On July 6 a flock of 30 birds flew east over Camp 1, and a flock of 90 was seen flying south over Camp 1 on July 7.

Specimens (7): ♂, 38922, testis 6 × 1 mm., 264 gm., breeding plumage; ♀, 38923, ova to 2 mm., 269 gm., breeding plumage; ♀, 38924, ova to 1 mm., 280 gm., 3 P old; ♀, 38925, ova to 1 mm., 319 gm.; ♂, 38921, testis 7 × 2 mm., 211 gm., breeding plumage; Camp 1, July 7. Male, 38927, fat light, 231 gm., 4 P old, Camp 2, July 9. Juvenile, sex not recorded, 38920, 43.0 gm., 1 mile south of Washington Beach, July 10. Two of our specimens, both males, are in worn breeding plumage and evidence no molt; another specimen, a female, is also in breeding plumage but is molting on the breast. The remaining two adult skins in our series are three-quarters through the molt and are for the most part in fresh winter feather.

Dresser (1866:37) took an unspecified number of specimens of the Willet at the "Boca Grande" in July and August, but actual breeding in Tamaulipas was first established by C. R. Robins, who found a "scattered colony of breeding Willets" and took a female with an egg in the oviduct on May 9, 1949, near Tepehuaje (Sutton, 1950:135). Sutton (*op. cit.*) has discussed the characters of this specimen and of birds from Cameron County, Texas. The specimen from Tepehuaje reportedly is closer to *C. s. inornatus* than to *C. s. semipalmatus* both in size and color, and birds from Cameron County are intermediate between the two subspecies in size but like *C. s. inornatus* in color.

TABLE 3.—MEASUREMENTS IN MILLIMETERS OF SPECIMENS OF CATOPTROPHORUS SEMIPALMATUS FROM THE BARRIER ISLAND OF TAMAULIPAS

Sex and Catalogue Number	Wing	Tail	Full culmen	Tarsus	Weight in grams
♂ 38921[A]	197	80.6	61.0	59.0	211
♂ 38922[A]	198	74.4	61.9	57.9	264
♂ 38927	194	75.5	60.4	56.4	231
♀ 38923[A]	201	71.0	59.0	55.4	269
♀ 38924	199	71.0	61.3	59.0	280

[A] Specimens in worn breeding plumage.

Measurements of our five adults from the barrier island are presented in Table 3 for comparison with those of *C. s. semipalmatus* and *C. s. inornatus* given by Ridgway (1919:316-319). Like the specimens from Cameron County examined by Sutton (*op. cit.*), our birds are intermediate in size between average-sized individuals of the two named subspecies. In color and pattern, we find that our specimens in breeding plumage fall within the range of variation of *C. s. semipalmatus* as exemplified by five specimens in nearly identical states of wear and fading in the Museum of Natural History.

On the basis of the evidence presently available, we are reluctant to follow Sutton (1950:136) in assigning breeding birds from the Gulf coastal region to *C. s. inornatus*, a name otherwise applied to a population of birds breeding inland, in northwestern North America south to central Utah and Colorado and east to South Dakota (and formerly to western and southeastern Minnesota and Iowa; see A.O.U. Check-list, 1957:190). The intermediate characters of birds breeding in coastal Texas and Tamaulipas probably represent not the results of actual genetic intermixing of the two named populations but, rather, an adaptive response of the eastern coastal stock (*C. s. semipalmatus*) to environmental modalities distinct from those operating elsewhere within the range of the eastern coastal population or on the inland population. Accordingly, we tentatively use the name *C. s. semipalmatus* for our Tamaulipan specimens, realizing that the patterns of geographic variation in the species do not lend themselves well to taxonomic treatment by the trinomial nomenclatural system. The need for a comprehensive analysis of geographic variation in this species, based, if possible, on proper segregation of age classes along the lines followed by Pitelka (1950) for *Limnodromus*, is obviously indicated.

PLATE 5

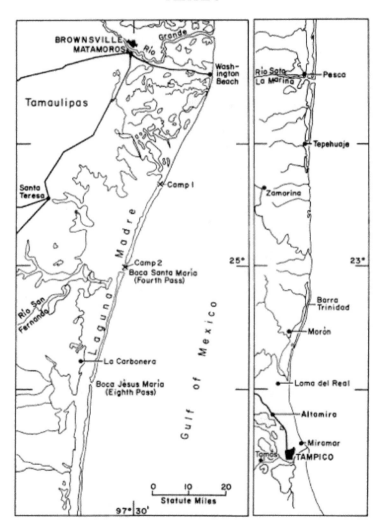

Map of coastal Tamaulipas, showing the barrier island and localities mentioned in text. Stippled areas are extensively marshy.

PLATE 6

Fig. 1.—Croton and Fimbristylis on stabilized dunes; the Laguna Madre and surrounding alkaline flats and clay dunes are visible in the background. Habitat of Road-runner, Ord kangaroo rat, and keeled lizard.

Fig. 2.—Active dune near Camp 1. Other active dunes can be seen in the background, in the right foreground is a clump of Croton, and in the left foreground is a small clump of Fimbristylis. Habitat of Road-runner, Ord kangaroo rat, and keeled lizard.

Arenaria interpres morinella (Linnaeus): Turnstone.—Approximately 40 individuals were noted along the beach from July 6 to 10, mostly in small groups; the largest flock included 15 individuals. Specimens (5): ♂, 38931, testis 4 × 1 mm., moderately fat, 107 gm., 4 P old; ♂, 38932, testis 3 × 1 mm., moderately fat, 103 gm., molting; 75 miles south of Washington Beach, July 8. Male, 38928, testis 2 mm., 111 gm., 3 P old; ♂, 38929, testis 3 mm., moderately fat, 106 gm., 6 P old; ♂, 38930, testis 2.5 mm., moderately fat, 108 gm., 6 P old; Camp 2, July 9.

The only previous record of the Turnstone in Tamaulipas is an observation of an unspecified number at Tepehuaje on May 9, 1949 (Robins, Martin, and Heed, 1951). The dates of our records suggest that nonbreeding birds summer along the coast of Tamaulipas. The species is present in small numbers in summer along the central Gulf coast of Texas (Hagar and Packard, 1952:8). Loetscher (1955:26-27) does not report records for Veracruz in summer, but records of the species in Yucatán on May 31, 1952 (Paynter, 1955:101), and on June 16, 1900 (Mexican Check-list, 1950:79), probably represent summering nonbreeders. Probably also in the same class are supposed "migrants" seen at Coatzacoalcos on May 17, 1954, and June 4, 1955 (Coffey, 1960:290).

Inasmuch as Haverschmidt (1955:368) reports that nonbreeding birds summering in Surinam only occasionally assume breeding plumage, it is noteworthy that our specimens were molting from nuptial (summer) to winter plumage. None of the nonbreeding northern shorebirds observed by Eisenmann (1951:183) in Panamá in summer were in nuptial plumage.

Crocethia alba (Pallas): Sanderling.—This sandpiper was noted each day along the beach, occasionally singly but more frequently in groups ranging from 10 to 50 individuals. Specimens (7): ♂, 38936, testis 2 mm., light fat, 49 gm., 5 P old, Camp 1, July 7. Female, 38937, ova to 1 mm., fat, 58 gm., 4 P old; ♂, 38939, fat, no weight recorded, 6 P old, breeding plumage; 3 ♂ ♂, 38940-38942, fat, no weight recorded, 4-5 P old; Camp 2, July 9.

With one exception as noted, our specimens are in worn, nonbreeding plumage and are replacing their old feathers with new ones fundamentally the same in color and pattern; the exceptional specimen is molting from worn breeding plumage into nonbreeding plumage. Only one other individual in breeding feather was seen on the island.

According to the Mexican Check-list (1950:99), the Sanderling has been recorded in México from August to May 19. In Texas, Peterson (1960:107) reports that it is a migrant, April to June and July to November, and that it winters along the coast. We suspect that many of the birds present in Texas in June and July, together with those recorded by us in Tamaulipas in July, are nonbreeding, summering individuals. Haverschmidt (1955:368) reports northern-summer records from Surinam, and, according to the A.O.U. Check-list (1957:208), nonbreeding birds occur in summer extensively through winter range of the species, including the Gulf coast of the United States.

Micropalama himantopus (Bonaparte): Stilt Sandpiper.—Two birds in worn winter plumage were taken as they foraged together at the edge of the laguna near Camp 2 on July 9. Specimens (2): ♂, 38934, testis 2.5 mm., heavy fat, 116 gm., 4 P old; ♂, 38935, testis 3 mm., fat, 111 gm., 4 P old.

Our specimens probably were nonbreeding birds summering between the breeding range in arctic America and the winter range in northern South America. The A.O.U. Check-list (1957:202) does not mention nonbreeding, summering records of this species. The 251 birds seen by Coffey (1960:292) at Cacalilao, Veracruz, on May 11, 1954, were probably migrants.

Recurvirostra americana Gmelin: American Avocet.—This species was seen only in three large flocks flying south along the beach, as follows: 56 birds 72 miles south of Washington Beach, July 8; 38 birds 73 miles south of Washington Beach, July 8; 29 birds 72 miles south of Washington Beach, July 10. All birds were in winter plumage.

All these birds were possibly autumnal migrants, but the dates are early; the species has not previously been recorded on migration in México before August (Mexican Check-list, 1950:101). The species is known to breed in San Luis Potosí (Mexican Check-list, *loc. cit.*) and along the lower coast of Texas ("rarely to Brownsville"; A.O.U. Check-list, 1957:209); avocets thus may also breed in coastal Tamaulipas.

Larus argentatus Pontoppidan: Herring Gull.—A first-year bird was observed near Camp 2 on July 8, and two subadult individuals were seen on the beach between the Third and Fourth passes on July 8.

Larus atricilla Linnaeus: Laughing Gull.—This gull was common all along the beach. Many individuals were in full breeding feather and many subadult birds were also present. Specimens (6): ♂ subadult, 38944, testis 5 × 1 mm., 325 gm., molting; ♀, 38945, ovary small, 309 gm., in molt, brood patches refeathering; sex?, 38943, 315 gm., in molt; sex? subadult, 38946, 327 gm., in molt; Camp 1, July 7. Female subadult (second-year), 38947, 305 gm., in molt, Camp 2, July 8. Female, 38926, ova to 2.5 mm., 313 gm., 8 P old, Camp 2, July 10.

The Mexican Check-list (1950:105) refers to the Laughing Gull as a common winter resident on both coasts of México from August 7 to May 17, but Loetscher (1955:29) found it locally common throughout the year on the coast of Veracruz, and he mentioned seeing birds a short distance south of Tampico in June and July. The status of this gull in Tamaulipas remains to be determined; probably it will be found breeding locally, but many of the birds summering in eastern México are most likely nonbreeders (A.O.U. Check-list, 1957:226).

Chlidonias niger surinamensis (Gmelin): Black Tern.—On July 6, 7, 8, 9, and on the morning of July 10, we saw this species only occasionally, recording in total not more than 50 individuals. But, about noon on July 10, we observed at least 300 birds in compact flocks of about 50 individuals each between Washington Beach and a point about 9 miles south of that locality. Approximately one in ten birds seen was in breeding plumage, the rest being in winter or subadult plumages, which are indistinguishable in the field. Perhaps some of the birds seen were nonbreeding, summering individuals, but we presume that the large groups were southbound migrants, and we note that autumnal migrants appear in northern Veracruz as early as July 1 (Loetscher, 1955:30). On the central Gulf coast of Texas, Hagar and Packard (1952:9) indicate that an influx of birds occurs in the last week of July, and small numbers of birds, presumably nonbreeding individuals, are present along the Gulf coast throughout June and July. Dresser (1866:45) found this species to be "common at the Boca Grande during the summer."

Specimens (2): ♂, 38948, testis 6 mm., moderately fat, 68 gm., in breeding plumage, Camp 1, July 7. Female, 38949, ovary inactive, 49 gm., molt into winter feather almost complete, Camp 2, July 10.

Hydroprogne caspia (Pallas): Caspian Tern.—The only published record of the Caspian Tern in Tamaulipas is a report of one seen at Lomas del Real on November 20, 1956 (Coffey, 1960:260), but we found it moderately common all along the beach and at the margin of the laguna. It was frequently associated with the Royal Tern, which outnumbered it better than three to one (see Table 2). The species is resident and breeds along the coast of Texas, and it probably has similar status in Tamaulipas. However, in Veracruz it is known only as a winter visitant (Loetscher, 1955:30) and as a spring migrant (Coffey, 1960:293). Specimen: ♀, 38950, ova to 2 mm., moderately fat, weight not recorded, 5 P old, Camp 2, July 9.

Sterna hirundo hirundo Linnaeus: Common Tern.—We took a specimen (♂?, 38951, no fat, 165 gm.), 49 miles south of Washington Beach on July 8, and saw two others over the laguna at Camp 2 on July 9. Our specimen had nearly finished with molt and feather growth into adult winter plumage. The status of Common Terns in Tamaulipas is uncertain; our record, and records from Tamós on July 1, 1952, and June 12, 1953 (Loetscher, 1955:29), probably pertain to nonbreeding, summering birds. Yet, the species has bred on the Texas Gulf coast (A.O.U. Check-list, 1957:235), and it reasonably may be expected to nest in Tamaulipas. Coffey (1960:293) saw two individuals at Altamira on May 10, 1954.

Sterna forsteri Nuttall: Forster Tern.—Six were recorded near Camp 1 on July 7, and two were seen on the beach on July 6 and 10. The Mexican Checklist (1950:108) does not cite records for Tamaulipas, but the A.O.U. Checklist (1957:234) includes northern Tamaulipas within the breeding range. Evidence suggesting breeding of the species in extreme northern Veracruz is reported by Loetscher (1955:29) in the form of a female specimen with "ovary greatly enlarged" taken seven miles west of Tampico on May 30, 1947. In the same area the species also seems to spend the summer as a nonbreeder, for Loetscher (*loc. cit.*) saw 20, nearly all in nonbreeding plumage, on July 1, 1952.

Specimens (4): ♂, 38952, testis 4.5 mm., 150 gm., 8 P old; ♂, 38955, testis 2 mm., 138 gm., 2 P old; ♂, 38953, testis 5 × 1 mm., 142 gm., 5 P old; ♀, 38954, ova to 1 mm., 148 gm., 2 P old; Camp 1, July 7.

Sterna albifrons antillarum (Lesson): Least Tern.—The status of this species in Tamaulipas is uncertain, but there is reason to believe that it breeds, at least in small numbers. We found the species moderately common and generally flying about in twos, possibly mated pairs, near both camps and on the beach. Breeding is suggested by the large sizes of the testes of the two males collected and by the presence of brood patches on a female taken on July 6, but we have no direct evidence of nesting in Tamaulipas, and it should be noted that this species is known to spend the summer in nonbreeding condition at many places (A.O.U. Check-list, 1957:239). Loetscher (1955:30) suggests that the species may be found breeding in Veracruz and mentions a record of 15 seen at Miramar, Tamaulipas, on June 26, 1952. Dresser (1866:45) found it to be "abundant" at the "Boca Grande" in summer.

On July 10, we saw flocks of 15 to 20 individuals flying along the beach a few miles south of Washington Beach.

Specimens (4): ♂, 38958, testis 11 × 4 mm. (right testis 5 × 4 mm.), light fat, 45 gm., 6 P old; ♂, 38959, testis 11 × 4 mm. (right testis 7 × 4 mm.), light fat, 45 gm., 6 P old; ♀, 38956, ova to 2.5 mm., 42.5 gm., 6 P old, brood patches refeathering; Camp 1, July 6. Female, 38957, ova to 1 mm., 44 gm., Camp 1, July 7. This last specimen had essentially completed the autumnal molt into winter plumage, with only a few feathers remaining ensheathed basally.

Our specimens are referable to *S. a. antillarum*, being paler dorsally and slightly lighter gray on the hind-neck than specimens of *S. a. athalassos* from Kansas, with which they were compared.

Thalasseus maximus maximus (Boddaert): Royal Tern.—This species was common all along the beach, occurring for the most part in flocks of from ten to 50 individuals in association with Cabot Terns. Data on gonadal condition and brood patches of some of our specimens suggest that breeding occurs in coastal Tamaulipas, as previously reported by the Mexican Check-list (1950:110). Robins, Martin, and Heed (1951) report seeing one Royal Tern near Tepehuaje on May 9, 1949, and Dresser (1866:44) found the species "common at the Boca del Rio Grande during the summer."

Specimens (6): ♂, 38960, testis 9 × 4.5 mm., not fat, 484 gm., 6 P old, brood patches refeathering, 4 miles south of Washington Beach, July 6. Male, 38961, testis 7 × 3 mm., 455 gm., no brood patches, 8 miles south of Washington Beach, July 6. Male, 38962, testis 10 × 5 mm., 387 gm., brood patches refeathering; ♀, 38963, ova to 1 mm., 358 gm., 3 P old; ♀, 38964, ova to 3 mm., 389 gm., 8 P old; Camp 1, July 7. Female, 38994, ova to 2 mm., 536 gm., brood patches refeathering, Camp 2, July 10.

Thalasseus sandvicensis acuflavidus (Cabot): Cabot Tern.—This tern was moderately common along the beach and margin of the laguna, and it was seen frequently in company with Royal Terns. Like the latter, this tern breeds in coastal Texas (A.O.U. Check-list, 1957:241), and it probably also nests in Tamaulipas, although direct evidence is not available. The only previous record of this species in Tamaulipas is a report (Robins, Martin, and Heed, 1951) of two observed on the beach near Tepehuaje on May 9, 1949.

Specimens (4): ♂, 38965, testis 9 × 4.5 mm., 208 gm., 9 P old, 49 miles south of Washington Beach, July 8. Male, 38966, testis 8 × 3 mm., not fat, 192 gm., 8 P old; ♀, 38967, ova to 3 mm., 193 gm., 7 P old, brood patches refeathering; ♀, 38968, ova to 1 mm., 186 gm., 8 P old, no brood patches; 52 miles south of Washington Beach, July 8.

Rynchops nigra nigra Linnaeus: Black Skimmer.—We found this species moderately common at the edge of the laguna at both camps and occasionally saw it along the beach. Generally two birds, probably mated pairs, were seen together; twice birds were seen carrying food in their bills, presumably intended for nestlings. The species is known to nest in Tamaulipas from "Matamoros Lagoon" south to Tampico (Mexican Check-list, 1950:112).

Specimens (2): ♂, 38970, testis 40 × 23 mm. (abnormally large, possibly as a result of hemorrhage), 418 gm., brood patches refeathering; ♂, 38969, testis 17 × 4 mm., fat light, 442 gm., brood patches refeathering; Camp 1, July 7.

Zenaidura macroura Linnaeus: Mourning Dove.—Our only record is a lone bird seen in a mesquite near Camp 1 on July 6. Possibly the species breeds along the margin of the laguna, although Aldrich and Duvall (1958:113, map) do not include coastal Tamaulipas in the known breeding range. Loetscher (1955:30) suggests that the Mourning Dove may be found breeding in the lowlands of northern Veracruz and cites a record of one seen at Tamós on July 1, 1952.

Geococcyx californianus (Lesson): Road-runner.—At least four individuals were seen in large dunes at Camp 1 on July 7 and 8. On several occasions we watched them pursue lizards (*Holbrookia propinqua*) at the margins of clumps of *Croton* and *Ipomoea*.

Chordeiles minor aserriensis Cherrie: Nighthawk.—Nighthawks of this species were seen regularly at Camp 1, where we flushed them from alkaline flats in the day and heard them calling as they foraged over the dunes in late afternoon.

Specimens (3): ♂, 38971, testis 5 mm., no fat, 62 gm., Camp 1, July 6. Male, 38972, testis 7.5 mm., no fat, 58 gm.; ♂, 38973, testis ?, no fat, 53 gm.; Camp 1, July 7. The gonads of these birds were not in full breeding condition, but it is highly probable that the birds were members of a population that had bred in the area.

Variation in *Chordeiles minor* in Tamaulipas has recently been studied by Graber (1955). Two specimens taken by him on August 3, 1953, approximately 9 miles south of Carbonera, resemble birds from Terrell County, Texas, and represent *C. m. aserriensis*, as do our three birds from the barrier island. Two of Graber's specimens from Lomas del Real, in southeastern Tamaulipas, are distinctly darker and probably represent *C. m. neotropicalis*, a subspecies subsequently described from Chiapas (Selander and Alvarez del Toro, 1955).

Muscivora forficata (Gmelin): Scissor-tailed Flycatcher.—On July 7 near Camp 1, two individuals were found in stands of mesquite. One was taken and proved to be an adult male (38974, testis 6 × 3 mm., not fat, 40 gm.) in postnuptial molt (6 P old).

We presume that the two birds recorded by us were members of a population breeding on the barrier island, rather than autumnal migrants. The Mexican Check-list (1957:69) records this species in México only as a transient and winter visitant. But, on the basis of records of birds seen along the highway between Matamoros and Ciudad Victoria, Davis (1950) has suggested that the species breeds in Tamaulipas, and this is supported by a report of one seen at the north end of the Monterrey Airport on June 1, 1957 (Coffey, 1960:294). Brown (1958) has recently established that the species breeds in Nuevo León by finding a nest 33 kilometers (by road) north of Sabinas, Hidalgo, on July 19, 1954.

Myiarchus cinerascens cinerascens (Lawrence): Ash-throated Flycatcher.—A juvenal male (38975, testis 2 mm., no fat, 35.0 gm.) taken in mesquite at Camp 1 constitutes our only record for this species. Lanyon (1961:441, map) has shown that most of Tamaulipas is devoid of these flycatchers in the breeding season; the nearest known breeding Ash-throated Flycatchers are slightly west of Corpus Christi, Texas, about 200 miles north-northwest of Camp 1 on the barrier beach. Our specimen closely resembles eight specimens from Coahuila, México, in general coloration and, especially, in the pattern of colors on the outer rectrices. Probably No. 38975 was from southwestern Texas or Coahuila and had begun its southward migration. Against this idea lies chiefly the fact that young-of-the-year tend to move south later than adults of the same species; so, this bird possibly had been reared in coastal Tamaulipas.

Eremophila alpestris giraudi (Henshaw): Horned Lark.—This species occurred in moderate numbers on alkaline flats and almost barren sand flats at both camps. At the time of our visit to the island, the breeding season apparently was coming to an end, but we noted no tendency in the birds to flock.

Specimens (7): ♂, 38981, testis 6 mm., 21.0 gm.; ♂, 38977, testis 7.5 × 4 mm., not fat, 27.5 gm.; ♂, 38979, testis 11 × 7 mm., 29.0 gm.; ♀, 38976, ova to 3 mm., brood patch vascular but regressing, no fat, 24.4 gm.; sex? juv., 38987, no fat, 21.0 gm.; sex? juv., 38980, 24.0 gm.; Camp 1, July 7. Male, 38982, testis 9.5 × 6 mm., 27.5 gm., Camp 2, July 9.

The subspecies *E. a. giraudi*, which is endemic to the Gulf coastal plain of Texas and Tamaulipas, has been reported in Tamaulipas previously only from Bagdad, near Matamoros (Mexican Check-list, 1957:106). The fact that our specimens show characters totally consistent with those of *E. a. giraudi* indicates that there is little genetic interchange between the population we sampled and those of *E. a. diaphora*, the closest of which reportedly breeds at Miquihana, in southwestern Tamaulipas.

Corvus cryptoleucus Couch: White-necked Raven.—Several groups of six to ten birds were present at Washington Beach on July 6 and 10; but, southward on the island, we recorded this species only once, on July 9, when a lone individual flew near Camp 2, being pursued and "buzzed" by two Least Terns. The Mexican Crow (*Corvus imparatus*) reportedly is common in the coastal region of Tamaulipas (Mexican Check-list, 1957:118) but was not seen by us.

Thryomanes bewickii cryptus Oberholser: Bewick Wren.—This species seemingly breeds in small numbers in mesquite stands near Camp 1, where we obtained a juvenile and saw another individual. Specimen: ♀ juv., 38983, no fat, 10.0 gm., Camp 1, July 8. *T. b. cryptus* is reported to intergrade with *T. b. murinus* of Veracruz in southern Tamaulipas (Mexican Check-list, 1957:160-161).

Mimus polyglottos leucopterus (Vigors): Northern Mockingbird.—We recorded this species only near Camp 1, where a few pairs were breeding in stands of mesquite. Males were in full song and territorial display.

Specimens (2): ♂, 38985, testis 11 × 7 mm., not fat, 43 gm.; ♀, 38984, ova to 4.5 mm., vascular brood patch, 49.0 gm.; Camp 1, July 7.

PLATE 7

Fig. 1.—Mesquite-cactus formation on clay dune at margin of the Laguna Madre west of Camp 1. Habitat of Northern Mockingbird, Cardinal, Bob-white, black-tailed jackrabbit, and Great Plains woodrat.

Fig. 2.—Batis-Monanthochloë formation on alkaline flats near the Laguna Madre, with mesquite bordering stabilized dunes in the left background. Salicornia, a classical dominant of salt marshes, is here relatively inconspicuous. Habitat of Nighthawk and Horned Lark.

PLATE 8

"Fossilized" burrow of Texas Pocket Gopher in a sandy trough between active dunes. A part of the cast has been broken away to show the general shape of the old burrow. The diameter of the cast is about 3.5 inches.

Cassidix mexicanus prosopidicola Lowery: Great-tailed Grackle.—Small, postbreeding flocks composed of both adult and juvenal birds were seen moving along the edge of the laguna at Camp 1. In the morning the flocks flew south, and in the afternoon groups of similar size flew north, presumably to a roost at an undetermined distance north of our camp. Occasionally, a few birds stopped to rest or to forage on the dunes or in stands of mesquite. At Camp 2 on July 9, a postbreeding adult female and a well-grown, presumably independent juvenile were taken as they perched in a clump of mesquite in which we found three old nests of *Cassidix*; two of the nests were about four feet apart in one tree, and the third was in another tree 100 feet from the first.

Specimens (4): ♂ adult, 38988, testis 6 mm., no fat, 209 gm., 6 P old, Camp 1, July 7. Female, 38989, ova to 3 mm., fat, 115 gm., old brood patch, Camp 1, July 8. Female, 38990, ova to 1 mm., moderate fat, 107 gm., 7 P old, brood patch refeathering; ♂ juv., 38991, testis 3 × 1 mm., not fat, 172 gm., 6 P old; Camp 2, July 9.

TABLE 4.—MEASUREMENTS IN MILLIMETERS OF ADULT MALES OF CASSIDIX MEXICANUS

LOCALITY	No.	Wing	Tail	Tarsus	Weight in grams
Austin, Texas	17-137[1]	184.3 (173-200)	203.8 (178-232)	46.38 (41.8-50.0)	225.6 June (204-253) 202.2 July (195-207)
San Patricio Co., Texas[2]	5	185.2 (182-188)	204.2 (190-219)	46.74 (45.1-50.2)	237.6 (228-245)
Barrier Is., Tamps.	1	178	185	47.1	209
Victoria, Tamps.[3]	4	192.2 (186-200)	224.2 (215-232)	47.77 (46.0-49.1)	254.3 (239-276)
Tampico, Tamps.[4]	1	197	214	48.3	260
Catemaco, Veracruz[5]	1	193	216	48.2	257

[1] Data from Selander (1958: 370, 373). Sample sizes, as follows: wing, 137; tail, 119; bill length, 20 (June and July); tarsus, 133; weight, 17 for June, 3 for July.

[2] June 13, 1961; breeding condition.

[3] May 6, 1961; breeding condition.

[4] May 7, 1961; breeding condition.

[5] November 28, 1959.

Specimens from the barrier island are clearly referable to *C. m. prosopidicola*, showing no approach to the larger and, in the female, darker *C. m. mexicanus* of Veracruz and San Luis Potosí. In Table 4, measurements of the adult male from the barrier island may be compared with those of specimens of *C. m. prosopidicola* from Texas and a specimen of *C. m. mexicanus* from Veracruz; it is apparent that our specimen is assignable to the former.

Evidence of intergradation between the two subspecies is shown in a series of birds collected near Ciudad Victoria, Tamaulipas, in May, 1961. The females in the series are highly variable in color individually, but are on the average paler than *C. m. mexicanus* from Veracruz; the males are distinctly larger than *C. m. prosopidicola* from Texas. At Miramar, near Tampico, Tamaulipas, a decided approach to *C. m. mexicanus* is also evident in the dark color of females and in the large size of both males (Table 4) and females.

Agelaius phoeniceus megapotamus Oberholser: Red-winged Blackbird.—This species was recorded only at Camp 1 on July 7, when we saw two males, one of which was flying south along the edge of the dunes in a flock of five Great-tailed Grackles. Specimen: ♂, 38992, testis 10 × 7 mm., fat, 54 gm., Camp 1, July 7. The large size of the testes of this individual indicates breeding condition.

Sturnella magna hoopesi Stone: Eastern Meadowlark.—Meadowlarks were found in small numbers along the margins of the alkaline flats at both camps. Breeding was still in progress, for males were singing and a female shot on July 9 had only recently laid eggs. Specimens (2): ♂, 38986, testis 13 × 8 mm., not fat, 102 gm.; ♀, 38987, ova to 6 mm., 3 collapsed follicles, not fat, 88 gm.; Camp 2, July 9.

Richmondena cardinalis canicaudus Chapman: Cardinal.—This species was recorded only in stands of mesquite near Camp 1, as follows: July 7, two pairs seen, from which a breeding female was taken; July 8, three birds seen. Specimen: ♀, 38933, edematous brood patch, 36.5 gm., Camp 1, July 7. Intergrades between the present subspecies and *R. c. coccinea* of Veracruz are reported from Altamira, Tamaulipas (Mexican Check-list, 1957:329).

Mammals

Dasypus novemcinctus mexicanus Peters: Nine-banded Armadillo.—Remains of an armadillo (89017) were found in a mesquite thicket in the dunes near Camp 1 on July 7. The bones are not badly weathered and were not embedded in sand.

This species has not been recorded previously on the barrier island of Tamaulipas, nor, for that matter, on any of the barrier islands on the western shore of the Gulf of Mexico.

Lepus californicus merriami Mearns: Black-tailed Jackrabbit.—From two to four individuals were recorded daily in dunes and on alkaline flats in the vicinity of stands of mesquite and cactus.

Specimens (2): ♀ adult, 89018, pregnant (two embryos, 28 mm. in crown-rump length), Camp 1, July 6. Male immature, 89019, Camp 1, July 7. Our specimens have been compared with two skins of *L. c. curti* from the type locality at Eighth Pass, with which they agree reasonably well in color. The size of the adult female is about that characteristic of other specimens of adult *L. c. curti*, but characters of the skull are consistent with those of *L. c. merriami*.

A specimen of this species from Matamoros and several from Brownsville, Texas, have been assigned by Hall (1951:43) to *L. c. merriami*. Specimens from Padre Island, Texas, reportedly resemble *L. c. curti* in smallness of the tympanic bullae but are in other characters referable to *L. c. merriami* (Hall, 1951:44).

Spermophilus spilosoma annectens (Merriam): Spotted Ground Squirrel.—These squirrels were moderately common in dunes at both camps. They were heard calling, and many tracks and holes were seen. On July 7, at Camp 1, a lactating, adult female (89020) and two dependent juveniles (89021, skull only, 89022, skin and skull) were shot at the entrance of a burrow; the uterus of the adult showed six placental scars.

Our adult specimen has been compared with ten specimens obtained by Hall and von Wedel at Eighth Pass in March, 1950; ours differs from the ten in being paler and slightly larger. The pallor is perhaps attributable to seasonal variation, and the size (246-79-38-7; weight, 133 gm.) is within limits that would be expected in a larger series of the population sampled by Hall and von Wedel. Hall (1951:38) referred specimens of this squirrel from Eighth Pass to *S. s. annectens*.

Geomys personatus personatus True: Texas Pocket Gopher.—This pocket gopher was abundant on low, stabilized dunes on the barrier island from four to 73 miles south of Washington Beach. One of us (Wilks) made a trip down the beach on May 20 and 21, 1961, and collected specimens at localities four miles south and 33 miles south of Washington Beach; additional specimens were taken at both Camp 1 and Camp 2 from July 6 to 10. At these localities the gophers seemed to maintain population densities approximating those of *G. personatus* on Padre and Mustang islands on the Texan coast.

There is but one other record of the Texas Pocket Gopher from México. Goldman (1915) described *G. p. tropicalis* from Altamira on the basis of specimens collected in 1898. Since that time, the species has not been reported as occurring south of Cameron County, Texas (Kennerly, 1954), some 50 miles northwest of the closest station of occurrence of the gophers on the barrier beach of Tamaulipas.

Our specimens are slightly smaller than *G. p. personatus* and slightly larger than *G. p. megapotamus*, the subspecies of nearest geographic occurrence to the barrier island. The degree to which our specimens differ in other respects, such as configuration of the pterygoid, is being studied further by Wilks. For the present, reference of our material to the nominate subspecies best expresses the relationships of these coastal gophers.

The fact that pocket gophers from the Tamaulipan barrier island occupy a position geographically intermediate between present Texan populations and the isolated population in southern Tamaulipas (*G. p. tropicalis*) helps explain the origin of the latter. It is likely that *G. p. tropicalis* represents the southern remnant of a once continuously-distributed population of pocket gophers living in coastal Tamaulipas in mid-Wisconsin to late Wisconsin time. At that time, sea level is thought to have been considerably lower than at present, exposing a sandy strip 80 to 100 miles wide off the present coastline. Presumably this would have been an area suitable for gophers and for southward dispersal of individuals from Texas. The only conceivable barrier to dispersal, and thus to a panmictic population, would have been the Rio Grande, but over the wide, low and sandy coastal plain the river channel almost certainly shifted regularly, thus decreasing its effectiveness as a barrier to movement. With subsequent rise in sea level, the gophers at Altamira became isolated and have presumably remained so for a considerable time. To judge by the marked morphologic differentiation of *G. p. tropicalis*, its degree of isolation from other populations has been much greater than those of populations inhabiting the Tamaulipan barrier island and the barrier islands of the coast of Texas. Contact between the latter two populations was probably fairly regular before man's stabilization of the channel of the lowermost reaches of the Rio Grande.

At Camp 1 we found evidence of the former occurrence of gophers in an area now largely covered by active beach dunes. Numerous skeletal parts of gophers and "fossilized" burrows (Plate 8) were found on the surface where troughs between active dunes reached down to an older, darker, and more tightly cemented layer of sand underlying the present dunes. It is clear that these gophers were not transported there, because the bones were not damaged, some of the skeletons were almost complete, and many of the bones were found near the "fossilized" burrows. Weathered but well preserved skeletal remains of at least 12 gophers were picked up at this site.

Specimens (17): ♀, 89023, Camp 1, May 20. 4 ♀ ♀, 89024-026, 89029; 3 ♂ ♂, 89027, 89028, 89030; Camp 1, May 21. Male, 89031, Camp 1, July 6. Three ♂ ♂, 89032, 89035, 89038; 4 ♀ ♀, 89033, 89034, 89036, 89037; Camp 2, July 9. Female, 89039, Camp 2, July 10.

Perognathus merriami merriami Allen: Merriam Pocket Mouse.—An individual taken in a trap in the dunes near Camp 2 constitutes the first record of this species from the barrier island of Tamaulipas. This pocket mouse seems to be uncommon on other barrier islands of the western Gulf of Mexico, for there is only one published report of its occurrence on Padre Island, Texas (Bailey, 1905:141). Other nearby stations of occurrence are Altamira, Tamaulipas (Hall and Kelson, 1960:477), Brownsville, Texas (Bailey, *loc. cit.*), and 17 miles northwest of Edinburg, Texas (Blair, 1952:240).

Specimen: sex?, 89040, skull only, Camp 2, July 10.

Dipodomys ordii parvabullatus Hall: Ord Kangaroo Rat.—We found this species uncommon and confined in distribution to dunes, in which it was recorded as follows: an adult female was shot and two other individuals were seen at night on July 6 at Camp 1; three were trapped near Camp 1 on July 7; two were trapped at Camp 2 on July 10.

Specimens (5): ♀, 89041, 2 placental scars, 46 gm., Camp 1, July 6. Male, 89042, testes scrotal, 47 gm.; ♂, 89044, 60 gm.; ♀, 89043, 44 gm.; Camp 1, July 7. Sex?, 89045, skel. only, Camp 2, July 10.

Our material does not differ significantly from specimens obtained by Hall and von Wedel at Boca Jésus María in March, 1950, which formed the basis for Hall's description (1951:41) of *D. o. parvabullatus*. This subspecies is presumably confined in distribution to the barrier island of Tamaulipas. Two immature specimens from Bagdad, Tamaulipas, were tentatively assigned by Hall (1951:41) to *D. o. compactus*, a subspecies known otherwise only from Padre Island, Texas.

Neotoma micropus micropus Baird: Southern Plains Woodrat.—This species was noted only near Camp 1, where numerous houses were seen in stands of mesquite and prickly-pear cactus and an adult male (89046, 330 gm.) was taken on July 6. This species has not been reported previously from the barrier island of Tamaulipas. Our specimen is referable to the nominate subspecies and shows no approach to *N. m. littoralis*, a subspecies known only from the type locality at Altamira, Tamaulipas (see map, Hall and Kelson, 1960:684).

Procyon lotor (Linnaeus): Raccoon.—A weathered skull and a broken humerus were found at Camp 2. The skull is being studied by Dr. E. L. Lundelius, who informs us that it matches a number of raccoon skulls found in archaeological sites along the Balcones Escarpment of Texas. Such skulls are larger than skulls of raccoons occurring today in Texas (*P. l. fuscipes*) and closely resemble skulls of raccoons (*P. l. excelsus*) presently confined in distribution to Idaho, eastern Oregon, and eastern Washington. Further details of this situation are to be reported elsewhere by Lundelius.

Taxidea taxus (Schreber): Badger.—Two burrows were found in the stabilized dunes near Camp 1, tracks were noted on the alkaline flats, and a weathered skull (89047) was found on the flats west of Camp 1 on July 7. The skull appears to be of an immature animal, for the sutures are not well closed and the teeth show little wear.

Our records require an extension of known range of this species southeasterly by approximately 50 miles. The only previous record in coastal Tamaulipas is based on two skulls from Matamoros (Schantz, 1949:301). The skull from the barrier island cannot be determined to subspecies but on geographic grounds is referable to *T. t. littoralis*, with type locality at Corpus Christi, Texas.

Canis sp.—Numerous tracks made either by Coyotes (*C. latrans* Say) or by domestic dogs were seen in dunes and on the beach at both camps. A weathered, posterior part of a canid skull was found in dunes at Camp 2 on July 10, and a partial left mandible was taken on the beach at Camp 1 on July 6. Unfortunately, specific identification of the skull fragments is not possible, but the few reasonably good characters that we can use suggest that our material is of domestic dogs rather than of Coyotes. Hall (1951:37) found tracks and other signs of Coyotes at Eighth Pass but did not take specimens.

Most of the canid scats examined by us contained remains of crabs and fishes.

Odocoileus virginianus (Boddaert): White-tailed Deer.—A weathered Recent fragment of a mandible (89048) and part of a femur (89049) of this species were found near Camp 1 on July 7, and a metapodal was picked up in the dunes at Camp 2 on July 9. This species has not been reported previously on the barrier island of Tamaulipas and it probably no longer occurs there, for we saw no tracks or other signs of it. Hall (1951) did not find it at Eighth Pass.

Our specimens probably pertain to *O. v. texanus* but are possibly of *O. v. veraecrucis*, which has been reported from Soto la Marina (Goldman and Kellogg, 1940:89).

The only species of mammal known from the barrier island of Tamaulipas that we did not find is the Hispid Cotton Rat (*Sigmodon hispidus*). Two specimens of this species trapped near Eighth Pass in March, 1950, formed the basis for the description of *S. h. solus* (Hall, 1951:42), a subspecies known only from the type locality.

Discussion

The known vertebrate fauna of the barrier island of Tamaulipas consists of one species of tortoise, two species of lizards, at least one (unidentified) species of snake, 49 species of birds (48 recorded by us and the Semipalmated Sandpiper), and 12 species of mammals. This is clearly a depauperate fauna, such as is characteristic of islands generally, and indicates that the peninsular nature of the northern part of the barrier island is of relatively small consequence in determining presence or absence of species. It is likely that the restricted environmental spectrum is much more important in this regard than is the fact of semi-isolation.

Of the 49 species of birds, 10 are known to breed on the island and an additional 21 are suspected of breeding either on the island or on small islets in the adjacent Laguna Madre de Tamaulipas. Eleven species occur on the island as nonbreeding summer residents, about which we will have more to say below. Four species have been recorded on the island in summer but breed elsewhere, that is to say, they only wander over the island (Man-o'-war Bird, Turkey Vulture, *etc.*). Two species are known only as migrants, and the status of one, the Sora Rail, is uncertain. The number of migrant species doubtless will be greatly increased by field work at those times when birds migrate.

The avifauna is not depauperate owing to the exclusion of any one of the three major zoogeographic stocks thought to be important in the development of the present North American avifauna (Mayr, 1946). If we examine the breeding passerine birds of the barrier island and the breeding passerine assemblage at the same latitude in lowland Sonora (Mayr, *loc. cit.*) as to their ultimate evolutionary sources, we find that for both places somewhat more than half the birds have developed from indigenous, North American stocks, about one-third have been derived from South American stocks, and one-fifth to one-eighth are from Eurasian stocks. It is most unlikely that such close correspondence in relative composition of the two avifaunas would occur by chance. Thus, we can only conclude that each of the historical avian stocks is proportionately restricted in numbers on the barrier island.

Faunistically, the barrier island resembles Padre and Mustang islands and the adjacent mainland of Tamaulipas and southern Texas, reflecting the relative uniformity of environment in this region. It is apparent that there is a faunal "break" or region of transition in the vicinity of Tampico, in extreme southeastern Tamaulipas. On the coastal plain, many tropical species and subspecies occurring in Veracruz are found north to Tampico but fail to extend farther northward to the barrier island of northeastern Tamaulipas. Axtell and Wasserman (1953:4-5), have already commented on this situation, mentioning a number of snakes and lizards that have differentiated subspecifically on opposing sides of the Tampican region. They also note that large numbers of the lowland Neotropical floral and faunal elements reach their northern limits of distribution within the zone of transition around Tampico, and, also, many Nearctic elements find their southern distributional limits there.

Our small samples of birds and reptiles from the island show no detectable morphological differentiation from adjacent populations. However, several of the mammals are moderately-well differentiated, but the patterns and degrees of geographic variation are such that we can only speculate on the historical derivation of the insular populations. *Lepus californicus curti* is presently known only from the barrier island of Tamaulipas, but Hall (1951:43) has suggested that it may also occur on the adjacent mainland. A resemblance between individuals of this subspecies and specimens of *L. c. merriami* from Padre Island in smallness of the tympanic bullae is regarded, probably correctly, by Hall (1951:44) as independent development—that is, parallel adaptation to similar environmental conditions reaching fullest expression on the barrier island of Tamaulipas. As is also true with *Geomys personatus* and *Neotoma micropus*, the barrier island population of *Lepus californicus* shows relationships with animals from Texas and northern Tamaulipas (*L. c. merriami*) and no connection with (resemblance to) animals from the south (*L. c. altamirae*, known only from the type locality at Altamira, near Tampico).

In color and cranial proportions, *Dipodomys ordii parvabullatus* of the barrier island is closer to *D. o. compactus* of Padre Island than to *D. o. sennetti* of southern Texas and the Tamaulipan mainland. But, *D. o. parvabullatus* resembles *D. o. sennetti* in external measurements (Hall, 1951:39). Possibly *D. o. parvabullatus* and *D. o. compactus* are phylogenetically closer to one another than is either to *D. o. sennetti*. It is also possible that each evolved independently from a mainland stock represented today by *D. o. sennetti*; the resemblance of the two insular populations would thus be a matter of convergence in response to like environmental conditions.

Sigmodon hispidus solus is an insular differentiate that probably reached the barrier island from the adjacent mainland of Tamaulipas, where its apparent closest relative, as judged by morphological similarity, now occurs.

Nonbreeding shorebirds in summer south of breeding ranges.—Certain aspects of this subject have already been discussed by Eisenmann (1951). As he notes, the phenomenon is more regular and widespread than generally has been appreciated. The old idea, that such oversummering individuals were "abnormal" or "senile," is totally inadequate, especially in view of the frequently large numbers of individuals involved.

Eisenmann's suggestion that nonbreeders are immature is probably valid, and it is supported by Pitelka's examination of dowitchers (1950:28, 51). For gulls, which can be aged by characters of plumage, there is no question that most nonbreeders are immature. Unfortunately, there are few criteria for determination of age in charadriiform birds.

With the possible exception of a specimen of *Limosa fedoa*, none of the presumed nonbreeding, oversummering shorebirds collected by us showed gonadal enlargement above expected minimal sizes for the species. Even so, the season was late at the time when we were on the island and most of the birds were molting; it is possible their gonads had been enlarged earlier in the season. Behle and Selander (1953) and Johnston (1956) have shown that nonbreeding first-, second-, and third-year California Gulls (*Larus californicus*) undergo gonadal enlargement in summer. Additionally, nonbreeding first-year males of certain passerine species (for example, the Brown Jay, *Psilorhinus morio*; Selander, 1959) are known to experience partial gonadal recrudescence in summer. It would be useful, and would facilitate discussion, to have data on gonadal condition of oversummering birds; any functional enlargement would be worth documenting.

Some species, notably the Semipalmated Sandpiper, Semipalmated Plover, and Black Tern, oversummer as nonbreeders in such large numbers that it is obvious that a significant fraction of the total population of the species does not breed in any one year. This raises questions concerning the possible ecologic situations that would select for delay in time of recruitment of young birds into the breeding segment of the population, assuming that nonbreeders are immature birds. Delay in maturation, or slow rates of maturation, may show general relationship to paucity of sites of breeding, as Orians (1961:308) suggests, but the shorebirds with which we are dealing breed in regions or in habitat-types not characteristically imposing general restriction on sites of nesting; more than one answer is necessary for the question even at this level. Data on age and numbers of nonbreeders, as well as on the ecology of breeding populations, are critical and are badly needed for most species.

In any event, species for which we have data demonstrating that they regularly oversummer south of their breeding ranges are probably adapted to having a part of their populations refrain from breeding each year. Whether this phenomenon can be explained solely in terms of selection at the level of individual birds (Lack, 1954) or involves selection of an adaptive response of the population as a whole (Wynne-Edwards, 1955; see also Taylor, 1961, concerning *Rattus*) is a problem that cannot be resolved at this time. We may note that the species involved ordinarily breed in arctic and subarctic regions, and it would seem advantageous (as set forth below) for nonbreeders to remain well south of such high latitudes. The numbers of oversummering individuals may fluctuate with over-all population density, possibly as a result of crude density, but possibly also as a result of emigration of individuals in excess of optimal density on breeding grounds (see Wynne-Edwards, 1959). One aspect of this phenomenon not explicitly discussed by Wynne-Edwards is the possibility that some individuals never move north to breeding grounds at all, perhaps as a result of a behavioral character genetically-grounded and mediated by delayed maturation of the neurohumoral "clock." This certainly would be an economical means by which population numbers could be regulated, for there would be a saving of energy in that some individuals not only would not move north, but also would not participate in the behavioral interactions involved in territorial spacing. Occurrence of these birds throughout southern North America, Middle America, and northern South America may thus reasonably be understood.

Milton Keynes UK
Ingram Content Group UK Ltd.
UKHW030839021124
450589UK00006B/671